chartered
management
institute

inspiring leaders

The leading organisation for professional management

As the champion of management, the Chartered Management Institute shapes and supports the managers of tomorrow. By sharing intelligent insights and setting standards in management development, the Institute helps to deliver results in a dynamic world.

Setting and raising standards

The Institute is a nationally accredited organisation, responsible for setting standards in management and recognising excellence through the award of professional qualifications.

Encouraging development, improving performance

The Institute has a vast range of development programmes, qualifications, information resources and career guidance to help managers and their organisations meet new challenges in a fast-changing environment.

Shaping opinion

With in-depth research and regular policy surveys of its 91,000 individual members and 520 corporate members, the Chartered Management Institute has a deep understanding of the key issues. Its view is informed, intelligent and respected.

For more information call 01536 204222 or visit www.managers.org.uk

■■■■ C O N T E N T S ■■■■

There is some trepidation that the top of this page may be out of date by the time we each reach the bottom it. We are living in such a fast-developing and fast-changing world now and M-Commerce is no haven from this; it is perhaps one of the fastest moving parts of our world.

Nevertheless, there is no need to fear M-Commerce. Much is yet to come in the next few years, but we have already seen it working today and have a clear idea of what may happen tomorrow. It is important to remember throughout that M-Commerce will not alter what we do, although it will affect how we do it. At both a personal and business level, the introduction of the Personal Computer 20 years ago has not changed what we do, but it has substantially changed how we do it. M-Commerce is no different. This book describes the main areas of M-Commerce and what it means for business and individuals.

If you or your business is not abreast of M-Commerce,

M-Commerce

- All things beginning with 'M-' for 'mobile'.
  M-Commerce, M-Business and the mobile internet
- The technology behind M-Commerce, including
  Personal Digital Assistants (PDAs) and
  mobile/cellular telephones
- The sort of content and applications M-Commerce
  brings
- The opportunities, risks and challenges that we
  should be prepared for

someone else will be and will seize the advantage that you
should be taking. If you could be reaping these benefits, read
on.

M-Commerce, M-Business and the mobile internet

All things mobile

Today we will not be talking about cars, bicycles and supermarket trolleys, but in general terms about what hs been described as 'The New Frontier of Digital Economy'.

It may seem sophistry to draw a distinction between M-Commerce and M-Business. However, each shares a distinct and new channel through which businesses can communicate with customers, other businesses, and their own employees.

- M-Commerce
- M-Business
- Mobile internet

M-Commerce is all about buying and selling services and products using *mobile* devices.

M-Business is about using these devices to improve the performance of the business itself. It is about enhancing the efficiency with which staff do their work, and improving existing and enabling new relationships between their customers and other businesses with which they work, such as external suppliers.

You are the reader of this book, but there have been many people and businesses involved in getting it into your hands. The first page will give you a clue: the commissioners, the writer, the publishers, the typesetters, the printers, and that does not include the organisations at the other end of the supply chain, such as the shop where you bought the book.

All parties have been involved in something 'M-' for this. My own last conversation with Hodder & Stoughton was on a mobile phone as I was crossing a railway station concourse. One day you may be able to read this as an e-book, downloaded from the internet to your PDA, using your mobile phone.

M-Commerce has resulted from the bringing together of many things, most of which existed already. New technologies and a recognition that we are all on the move are the main factors involved.

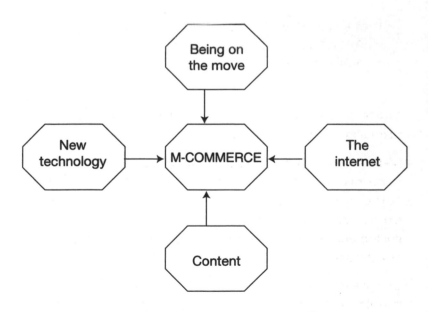

The cause of M-Commerce

Where is M-Commerce now?

Today is about where M-Commerce is at the moment. We will be discussing this and the future in more detail later in the week, but an overview is given below.

In the meantime, please bear two things in mind:

1 This book was written on a PDA and e-mailed, using a mobile phone, to the publishers from a train. Much work is done this way, by embracing what technology offers us today.

2 NTT DoCoMo is Japan's leading mobile phone operator. It has a vision of the future which supports the expansion of mobile devices beyond what is available now, to devices for people's pets, cars and many other things.

Expansion of mobile devices in Japan

	2010 projection (millions)
Humans	120
Cars	100
Bicycles	60
Portable PCs	50
Pets	20
Motorcycles, vending machines, etc.	10
Total	360

Source: DoCoMo Vision 2010, 2001.

Perhaps the table is visionary and aspirational, but DoCoMo already has more than 20 million subscribers.

M-Commerce is still in its infancy, yet it has the potential to be successful, hugely successful, simply because of the huge number of people who will be able to use it. It is more than what is available on the internet today, although it uses some of the same technology.

Today's technology

Already there is an abundance of mobile phones and PDAs; hardly a day passes without a manufacturer announcing a new device or new functionality. Unlike desktop PCs, these devices can be with people who are either potential customers or staff, throughout their day.

Mobile phones and hand-held computers are and will certainly be the pre-eminent devices involved in M-Commerce. Other devices will play an increasing role, especially in cars.

Today's mobile phone networks
Mobile phones have become more and more popular as the

quality of their networks has improved. Essentially, this has developed as shown in the table below.

Date	Type (G = generation)	Features
1979	1G	Analogue
1991–1999	2G	Digital
1999–2001	2.5G	Digital (Transitional)
2001–	3G	Digital
?	4G	Digital

Although each generation of mobile telephone systems has been based on a cellular architecture (hence the name 'cellphone'), there have been progressive improvements in the quality and reliability of their traffic.

The migration from analogue to digital networks has seen obvious benefits in terms of the quality of voice transmission but, significantly for M-Commerce, it has paved the way for more and better features to be made available to data traffic (computer-based as opposed to voice).

Perhaps the most obvious aspects of the second generation of mobile networks have been:

- Agreement of standards for cellphones amongst national regulators of the Global System for Mobile Communications (GSM); this is now widely deployed in most of the world, except for Japan and the US, and usage in the Americas is increasing.
- Introduction of the Short Messaging Service (SMS). A huge number of these are already passing over

GSM networks. Approximately 13 billion SMS messages are sent annually in Germany, and in the UK 6 per cent of GSM revenues come from SMS.

Today's M-Commerce applications

Enabling technology and networks are obviously important to M-Commerce, but success will lie in the content that is delivered.

This is not necessarily the same as the content available over the internet through fixed land-lines. The key to using M-Commerce is to recognise that 'M-' stands for mobile and that your customers and your staff are on the move.

You should identify what that means for them, understand what they need or want while they are on the move, and appreciate what they will be prepared to pay for in terms of the products and services you can deliver.

You must understand that what potential customers want while they are moving is not necessarily the same as when they are at their offices, factories or homes. Not only will they want different things, but they may also want the same things delivered to them differently, and usually quicker.

The following are already in use or their adoption is imminent:

- SMS
- Telephone (do not forget that actually talking to people remains important; many people prefer to talk with a real person than to a machine)
- E-mail

- Information delivery
- Mobile banking
- Voice mail
- Digital news
- Mobile audio (MP3)
- Access to the Worldwide Web
- Video phone and mail
- Education including receiving or having access to training and reference material
- Car navigation (knowing where you are and getting directions to where you want to be, including up-to-date information about traffic problems)
- Medical diagnosis, similar to that already available on the internet where you can present symptoms. It is obviously no substitute for consulting a physician in person
- Catalogue shopping, which allows you to buy products while on the move
- Mobile TV

These are very general applications and you will be able to think of more specific M-Commerce applications for your own industry. Of course, mobility and the ability of the mobile phone to know where it is situated, are keys to this.

An example of an industry-specific application is the ability of an airline passenger to check-in for their flight, using a mobile phone. Various companies, including United Airlines, Swissair and British Airways have tried variations of this. They do, however, remain constrained by security considerations.

We have covered how consumers use M-Commerce today. Equally important are the opportunities that M-Commerce offers businesses internally. Although we have focussed on consumers, corporations are already seeing the value of M-Commerce for their business.

For the mobile employees, whether 'professionals', sales or service staff, they can, at present, gain access to:

- SMS
- E-mail
- Calendar and diary management
- Contact management
- Information about the availability of parts, products or services

Customer Relationship Management

It is difficult to talk about M-Commerce and businesses as a whole without mentioning Customer Relationship Management (CRM). Clearly, it is a very hot topic at the moment and is addressed on Wednesday. It probably merits a book on its own. However, here we shall concentrate on M-CRM (see also Wednesday).

Many CRM implementations fail to deliver the expected benefits to the business for two main reasons:

- While CRM systems may contain a wealth of information for sales or field service staff, they need to be back at the office to access it.

- CRM systems are wholly dependent on customer-facing staff who maintain that wealth of information.

This is often not done because:
- Staff need to be in the office to do it
- It is too difficult
- There is no obvious benefit to the individual.

Use of mobile technology can address these difficulties and transform CRM applications into the effective and accessible tools they should be.

Summary

Today has given an overview of where the mobile world is now. Later this week, we shall cover more about the technologies themselves, the applications and content involved, and what M-Commerce means for individuals and for businesses.

The technology – mobile communications

We shall discuss mobile devices tomorrow, but today you will find out about the infrastructure, the networks, in which these devices can work. By way of warning, this is the most technical chapter of this book, but it is important that you know the basics of the underlying technology so that you can be aware of the opportunities it offers.

In the UK you may have heard of telephone companies who have spent huge amounts of money buying licences for third generation (3G) phones and networks. You may be asking yourself 'What are these generations?' and 'Why haven't we seen any real improvements yet?'

As we saw on Sunday, there has been a progression through newer generations of network capability. Each of these generations has been evolutionary, rather than revolutionary. The second (2G) and transitional (2.5G) generations are (or will be shortly) familiar to us. The coming third generation (3G) will, however, represent a significant change in what it offers us.

First generation (1G)

The first generation (1G), an analogue system, used frequency modulation (FM) as the transmission method. This generation originated the concept of 'cells' or 'cellphones'.

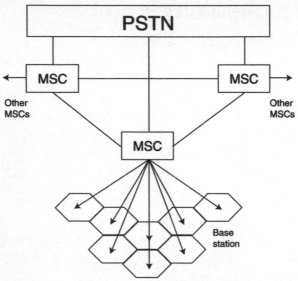

Each cell has a base station. Cells, when grouped together, form a cluster. The Mobile Switching Centre (MSC) is connected to all the base stations in a cluster. The MSC itself is connected to MSCs of other clusters and to the PSTN (Public Switched Telephone Network) switching centre. Adjacent cells use different frequencies, which means that non-adjacent cells can reuse frequencies.

It is convenient to think of the PSTN as the normal land-based telephone service. MSCs route traffic between cells. Each cell is a fixed area on land, inside which many people can use their cellphones.

First and second generation systems were deployed in the 800 Mhz band (Americas, Pacific Rim, Africa, Eastern Europe and Russia), and the 900 Mhz band (UK). Subsequently, frequencies were made available in the 1900 to 2100 Mhz range.

Second generation (2G)

The evolution from analogue to digital systems (i.e. the second generation or 2G) brought with it significant benefits:

- Encoding voice messages as digital bits which could be sent faster than they were spoken
- Compression meant that several voice messages could simultaneously use the same frequency
- Increased capacity meant that the size of cells could be made smaller or the number of adjoining cells increased
- Digital handsets require fewer components and less power – battery life is increased
- Error detection and correction is increased
- Authentication and encryption are more secure

In the 1980s, Europe had nine incompatible standards for cellular telephones. These were eventually brought together as the Global System for Mobile Communications (GSM). Other parts of the world have also struggled with competing and conflicting standards, but the Code Division Multiple Access (CDMA) is becoming pre-eminent in America and Asia. Although GSM and CDMA use very different technologies, they both support digital transmission and either circuit-switched or packet-switched data.

Note: circuit-switching is the traditional mode of telephone communication, which can be thought of as the presence of a channel, whereas packet-based refers to the ability to send traffic (either voice or data) in a number of discrete packets.

Transitional generation (2.5G)

2.5G is with us now. It is not considered a separate
'generation' in its own right, rather an upgrade of 2G. It is
still based on GSM (in Europe) or CDMA, but it is a
significant development. Characterised by existing networks
being upgraded to GPRS (General Packet Radio Service),
2.5G offers three major benefits:

- Speeds of up to 100 Kbits/second
- Always-on connectivity
- It is significantly cheaper to implement (certainly
  compared to 3G), and by and large it can be
  implemented with software changes only and no
  changes to the physical network infrastructure

Tuesday, Wednesday and Thursday will address what these
attributes of 2.5G mean for the development of physical
devices and, more importantly, for the commercial
applications that can be delivered.

Third generation (3G)

3G (sometimes referred to as UMTS – Universal Mobile
Telecommunications Services) should revolutionise mobile
communications in the same way that the internet has
revolutionised IT.

Third generation networks utilise improved technology and
give high bandwidth multimedia capable connections to
your mobile.

The British government has auctioned licences for the development of 3G mobile services. The UK and Japan have represented the first opportunities in the world to secure radio spectrum specifically for UMTS – the new global standard for mobile communications.

This should act as a springboard for UMTS opportunities across Eurasia and, potentially, the world.

The UK is an ideal 'point of entry' for UMTS in Europe; spectrum auctions and licensing processes are expected to take place in other European countries after the UK. This is largely due to the timing of the auction in the UK, but also the readiness of UK network operators to take up the opportunity which is in itself predicated on their perception of acceptance in the UK market and take-up generally of mobiles. The auction was an opportunity to exploit fixed/mobile convergence.

Access to mobile services and spectrum will soon become a business need for software, IT and even broadcasting houses.

How much was paid for 3G?

Licence bidder	Price
Hutchison	£4,384,700,000
Vodafone	£5,964,000,000
BT3G	£4,030,100,000
One2One	£4,003,600,000
Orange	£4,095,000,000

Why did they pay so much?
UMTS is the European element of the global IMT (International Mobile Telecommunications)-2000 standard for mobile communications systems. IMT-2000 could provide the

basis to integrate satellite, digital, cellular and cordless in a mobile device. UTRA (UMTS Terrestrial Radio Access) has been adopted in Europe as the radio interface standard, with an almost identical system in Japan.

Roaming (the ability to move from one network to another) across Europe, Japan and other Asian countries should be possible, with potential for US roaming, subject to development of a common standard by the standardisation organisations (ETSI, ARIB and T1P1).

Common standards are important for roaming between networks so that it really does not matter where the user is. From a wider perspective, to achieve a genuinely global implementation of the technology, and therefore commercial markets, it is important that everyone 'speaks the same language'. For technology, this means agreement on standards. There are clearly various interests involved in the development of standards, especially where it might involve protecting a previous investment.

In an ideal world, standards would be self-enforced by market pressures. Consequently, much of the work done by standardization organizations is less about monitoring and policing, but more about getting their agreement and developing them as the technology progresses.

3G is the evolutionary next step from 'second generation' digital systems and can be combined with existing GSM systems on dual-band terminals.

3G is forecast to revolutionise mobile communications in the same way that the internet is revolutionising IT.

3G should provide fast, secure and truly universal voice and data transmission (it provides up to 2Mbps [Megabytes per second]), breaking down the barriers between fixed and mobile telecommunications, broadcasting and information technology.

Hutchison Whampoa (see table of UK bidders for 3G) has announced plans to launch its 3G services in Britain in winter 2002, and after that in Italy and Hong Kong. It is likely to be the first 3G operator in Europe.

Fourth generation 4G
In March 2002 the International Telecommunications Union (ITU) met to throw some light on the fourth generation cellular standard. Unsurprisingly, it means many different things to different people. It is not yet a defined standard but more of a concept in which devices will be able to switch networks. At the moment carriers do not want it to be anything more than a concept. Given that they paid so much for 3G they would be much happier to wait 15 years while 3G pays for itself.

In essence, the 4G concept forecasts:

- Personal area networking, which will enable devices to communicate with each other; presently the dominant technology for this is Bluetooth
- The ability for devices to connect to high-speed access points
- Faster connection speeds (30 Megabytes per second by 2005 and 100 Megabytes per second by 2010)

What will this technology give us?

The technology itself gives us nothing. However, it does give an infrastructure upon which services can be delivered. More detail will be covered on Wednesday, but in the meantime UMTS is intended to support the following new high speed, interactive services for mobile phones, mobile terminals and laptops:

- Real-time high quality video conferencing
- Fast internet and corporate intranet access, regardless of location (Web page photo: UMTS = 0.4 secs; GSM = 83 secs; fixed line = 28 secs)
- Broadcasting and audio on demand
- Online banking and shopping
- Enhanced quality voice, fax and e-mail
- In-car real-time navigational systems
- A virtual office on one line with *simultaneous* voice and data services: make calls, receive faxes and remain on the office network simultaneously

Further development in the UK is undoubtedly on its way. Looking at what has already happened elsewhere in Europe gives us an insight into what is coming and what opportunities await us.

Mobile penetration

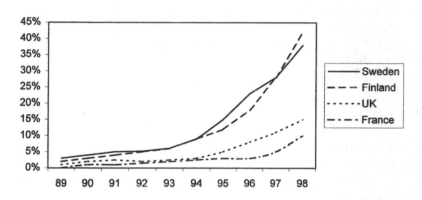

There has clearly been a high degree of substitution of fixed lines for mobile communication in Scandinavia (almost 100 per cent of men in their twenties), while penetration is less advanced elsewhere.

Today there are about 400 million subscribers around the world to wireless networks. This number is expected to grow to 1 billion by 2004, exceeding the number of physically wired phones.

Wireless networks are certainly attractive to implement in countries where it is difficult or too expensive to lay cables. They also mean that people are less separated by geography and more by time zones.

Summary

We have discussed fully the generations of vast wireless networks. Nevertheless, it is still worth remembering that the technology can be used, just as usefully, over short distances inside homes, offices and other buildings. This means that Local Area Networks (LANs) can be built in these environments without having to use wires and cables.

We have seen today how the networks for mobile devices have developed and continue to do so. The generations of these networks give us an infrastructure that opens commercial opportunities to us.

The technology – mobile devices

'My strong suspicion is that in the near term more people will lose data on their PDAs by running over them in their driveways than they will by viruses.'
Frank Price, senior analyst, Forrester Research

It is not just PDAs that we will be using. They are just one of a range of terminal devices that will connect, wirelessly, to the networks that we discussed on Monday.

- Two-way pagers
- Mobile phones
- PDAs
- WAP (Wireless Application Protocol) phones
- Visual phones
- Hand-held Personal Computers
- Devices inside machines (e.g. cars)

There has already been a great convergence of these devices as our mobile phones have more and more functions, even with 2G, and our PDAs communicate more. It can be confusing to consider these because they are frequently defined as different things by device manufacturers themselves. However, for our purposes, we shall consider a device to be a PDA, a mobile phone or a 'smart' device.

Personal Digital Assistants (PDAs) and Hand-held Personal Computers (HPCs)

PDAs have been used by businesses and individuals for a long time now, used initially rather like notebooks to store a diary and address book. This has developed through sending and receiving e-mail to running industry-specific applications, such as inventory management.

The popularity of PDAs has increased markedly over the past few years, both for individuals and for businesses. Partly because they have grown to be more socially acceptable, but our view of what they are for has also changed.

PDAs are no longer seen as only Personal Information Managers (PIMs), which was how they started, nor are they seen as replacements for desktop PCs. Aside from the industry-specific applications that have been developed for them, the key attributes are:

- The ability to synchronise data with systems at work
- Connectivity with the internet, typically for e-mail

Both of these are essential to the mobile worker. PDAs do not replace the office PC, but concentrate on the critical things that the mobile worker needs.

That PCs are everywhere has led to the social approval of PDAs. In addition, many applications developed for the PC have smoothly migrated to the PDA without any of the trauma associated with their initial introduction.

Having reduced this type of device to PDAs and HPCs, the range of products can still seem overwhelming. It is useful to think in terms of the Operating System used, i.e. EPOC, Palm OS or Windows® CE. Below are the things to consider when choosing a device for a particular type of user and usage.

Operating system
Symbian, the developer of EPOC, is a joint venture between Psion, Ericsson, Nokia and Matsushita. Although EPOC has been successfully used in Psion PDAs, for example, the Revo and Revo+, its future direction can be seen from Psion's partners in Symbian, i.e. mobile phone manufacturers.

Windows® CE is a scaled-down version of Microsoft Windows®, developed primarily for use on hand-held computers. Its implementations frequently come with similarly abridged versions of Microsoft applications, such as Word, Excel and Outlook.

Palm OS was developed specifically for palm-sized devices. It is thought to be very user-friendly and offers the essentials a user may be looking for. The Palm OS platform currently accounts for 70 per cent of the hand-held market.

Battery life
For a device to be truly mobile it must allow its owner to be away from a battery recharger (e.g. home or office) for a significant amount of time. Battery life depends on how the device is being used, whether for an occasional reference to an address book or for lengthy communications, be they wireless or cabled.

HPCs and PDAs typically use RAM, rather than disk, for data storage. The latter can be expensive in terms of battery life. It is worth considering what else uses a lot of power, such as back-lit or colour displays. Typically, battery life can vary between hours and days, but you can normally depend on a day's work before the battery needs recharging.

Size

In this instance, size is important. For a device to be truly mobile it must be small enough and light enough to be carried easily, in a breast pocket, purse or handbag.

Technology, in terms of microproccessors, continues to get smaller and now we are more constrained by the physiological constraints of the users of PDAs, such as the size of their fingers or their eyesight.

To reduce their size, many PDAs have moved away from emulating desktop PCs and completely dispensed with the traditional QWERTY keyboard. Instead, they have incorporated other input devices, such as handwriting recognition on touch-screens, or an ability to respond to voice commands.

Frequently, this has involved designing or modifying applications specifically for this. Sadly there remain constraints which even the genius of technologists cannot overcome. Speech recognition in noisy environments (e.g. factory floors or railway stations) is less than perfect, but no harder than talking to the person standing next to you.

Cost
PDAs come in a wide range of prices and they are becoming cheaper every day. This, in itself, is contributing to their more widespread use. They can be seen as a substitute for a full PC if only the very basic functions (e.g. PIM) are needed.

Applications
We will talk about the sorts of applications that are delivered on PDAs on Wednesday and Thursday. Suffice it to say that thousands of software programs are available and easily downloaded over the internet, ranging from business tools for personal productivity to games. A PDA will typically have:

- Diary
- Address and phone book
- Clock, with Alarm functions
- Note / memo pad
- Calculator
- Word processor
- Spreadsheet
- Database functions
- Games

Critical to these functions is the ability to synchronise with a central server or to download and upload documents and databases at either location.

Most PDAs will synchronise using infrared, cable or the internet. This ability is critical to the success of a PDA because the mobile user always needs to have the most up-to-date information.

Mobile phones

The mobile phone is the device that most clearly meets the needs of the mobile worker. Mobile phones may not have the functionality of a PDA, but this is changing, especially as we see convergence between these two types of devices.

Already, mobile phones offer the following features:

- WAP
- Short Message Service (SMS)
- Predictive text
- Clock and alarm
- Vibrate alert
- Picture editing
- Mobile chat
- Voice dial
- Voice memo
- Calendar or scheduler
- Phone book
- Games

Moreover, future phones will also have:

- Full QWERTY keyboard
- Screensavers
- MP3 playback where audio files (such as music) are compressed to reduce their size
- FM stereo radio
- Currency converter
- Blue Tooth compatibility, a short range (10 metres) wireless technology which allows enabled devices (such as domestic appliances and vending machines) to communicate with each other
- To-do list
- Television

GPS

Mobile phones are now beginning to have Global Positioning System (GPS) technology built into them. With this technology users can look at a map which shows their current location and, if the user gives the name of the place they want to get to, it will select the most appropriate route and guide them to their destination.

Phone services that allow users to see where they are on a map have been around for some time. However rather than using GPS satellites these earlier phones read signals from the phone networks and base stations located in the area to calculate where the user was. For that reason they could only give a rough approximation of the user's whereabouts and could be wrong by up to 500 metres. Phones incorporating GPS technology will be much more accurate with the user's location shown as a specific point on a map and they can be accurate to just a few metres.

There has been an increasing convergence of mobile phones with PDAs. This initially manifested itself as making the devices more 'fun' to use, but features become more relevant to the mobile professional every day.

The technology behind many of the 'fun' applications directly relates to more serious features. For example, the ability to display moving images not only allows you to watch TV, it also opens the door to video-conferencing and streaming. Business applications available on the converged devices include:

- Internet Web browser
- Mobile e-mail
- Word processor
- Spreadsheet
- Presentation viewer
- Mobile colour screen
- Infrared port
- Fax

PDAs are developing rapidly. The eduPAD is specifically designed for education, but it illustrates how devices can be devised for targeted users and purposes. Use of the eduPAD is being piloted by the Ministry of Education in Singapore. It has the basic functions of a PDA but it can also be used to access electronic publications stored in memory cards the size of postage stamps. Importantly, because it is portable (it weighs 800g and has a colour display of 18cm from one corner to its diagonal opposite) it can be used anytime and anywhere. In addition to memory cards, it has a microphone, speakers and an infrared capability which allows the classroom, teacher and pupils to interact.

This device costs less than $1,000 to produce each one. But it illustrates how businesses are not constrained by hardware in arriving at something that is specifically made to support their own commercial aspirations.

Synchronisation with the office becomes more and more important and more readily available. Apart from diaries and contact details, mobile workers can access up-to-date product information and stock availability.

Not only are networks progressing to a third generation, we are also seeing 3G handsets. 2.5G and 3G give much broader bandwidth and greatly increase the amount of information that can be passed to these new devices, and this allows applications such as video streaming. More importantly, is the 'always-on' capability and the implications for continuous and up-to-date synchronisation.

In terms of the variety of devices, organisations may find themselves supporting up to 50 different types of devices and networks. Remember that the value of mobile handsets is not solely in themselves, as stand-alone devices, but also in their ability to communicate, especially with the mobile professional's workplace.

Regardless of how much mobile devices develop, we remain constrained by our physiognomy. We are aready seeing greater use of mobile phones plugged into laptop computers. In doing so, the user can preserve all the facilities of the laptop (such as the larger screen) and enjoy the benefits of mobility afforded by the mobile phone.

This means that the workplace can now travel easily with the mobile worker. Of more relevance is the fact that the workplace can travel to the consumer. Tomorrow, we shall see the significance of this for commercial applications and, although important, mobile internet is only one of the mediums to consider.

What M-Commerce means for business

This book does not deal with the internet or E-Commerce specifically. These subjects are already covered in the *In a week* series. They arc, however, relevant to M-Commerce in terms of:

- The dramatic speed with which the internet has become established
- The lessons that a mobile world can learn from how the internet has been deployed
- The sheer dependence of many mobile applications upon the internet

We live in an E-enabled world and all our thoughts about a mobile world should be in that context. Much of M-Commerce is about providing on the move what the internet can already provide at our workplaces. Here we concentrate on the inherent differences, both in the technology and how we work, and on the opportunities and risks that M-Commerce presents.

Yes, M-Commerce is certainly a new medium in which we can operate. It will not fundamentally change much of what we do, but rather how we do it, hopefully for the better. Many businesses are trying to find new channels through which they can communicate with their customers and M-Commerce offers a new opportunity.

Mobile devices are becoming more and more popular with individuals. We will explore this, and how individuals are affected both as consumers and workers, on Thursday. Today, our focus is on businesses and how they can use much the same technology to give themselves a new channel; both for direct communication with customers and to improve their internal processes and customer services. Nevertheless, it is important to remember that 'M-' is an additional channel, and not necessarily a replacement for existing ones, such as the internet.

Most businesses aim to make money, although some may have less pecuniary aspirations, but the ability of workers in any organisation to be mobile, rather than chained to a desk or factory floor can only be for the better. We will look at privacy implications later. As well as having mobile access to diaries and e-mail, workers still need to be able to see and update corporate systems.

Field workers use the corporate systems before they go out, but once they have left their base they typically suffer from:

- A lack of immediacy in getting corporate information
- Difficulty in updating home systems when they need to (i.e. when they are actually with customers)

Customer Relationship Management

The mobile Customer Relationship Management (M-CRM) market is forecast to grow from $75 million in 2000 to $1.7 billion by 2005.

Implementing traditional CRM systems has been hard enough in itself. Much of the reason behind this is because the people who can update the systems do not have adequate or timely access to them. M-CRM goes some way to address the difficulties involved.

CRM is aimed at attracting customers, keeping them, decreasing churn and identifying further opportunities to sell. Overall, M-CRM will speed up the sales cycle. Many of the transactions normally involved in the most simple sale can be instant.

Likewise, other components of the relationship with the customer can be instant. For example, knowledge of:

- Previous contacts
- Previous purchases
- Requests for information
- Problems and complaints raised

The visiting field officer can give the customer up-to-date information about:

- The status of current orders
- Corporate news
- New product information

This may not be on the agenda for the meeting itself, but it is useful information for pre- and post-meeting conversations which remain vital in building and fostering the important personal relationship between field officer and customer executive.

The key to much of this is:

- The immediacy with which information can be relayed to customers
- The ease and speed with which field officers can personally update corporate systems with information about their clients
- The confidence that field officers can have in their information.

CRM application vendors are rapidly developing their systems to be usable by the mobile workforce, particularly where they have already developed versions of their systems usable on a notebook computer. It is not an enormous leap to transfer them to PDAs and use wireless, rather than wired, communication. They include the well-established CRM systems providers, such as Siebel and Oracle, but also new developers who are using M-CRM as a platform to deliver speech-enabled systems.

CRM, and especially M-CRM, initiatives must not be seen as purely technological solutions to wider problems. Sales and customer relations staff and field technicians may be mobile and have access to corporate data wherever they are working, but it is the people and processes that make it work, more so than the technology.

Fostering the relationship with the customer is not simply about knowing what the customer wants, but actually delivering it. Here, again, features of the M-Commerce world can help. First, a word of warning about managing expectations.

Expectations

The internet, fast corporate networks and aggressive marketing by device and network providers have all served to raise expectations of the mobile world in the minds of consumers and workers.

Mobile devices have to be:

Trying to achieve this on devices which must remain small

- Secure
- Reliable
- Manageable
- Scalable (capable of being adapted to reflect any changes in scale, which may be in any dimension)
- Robust
- Highly available

and light is a very tall order indeed. Expectations have been raised so much that people now expect to be able to use their mobile devices in the same way they use their desktop PCs. It is worth saying again that mobile phones and PDAs are not substitutes for desktops. They are intended to be used in different ways for different things.

As organisations introduce these devices and mobile applications, they must be ready to manage high expectations. This is particularly the case where the most enthusiastic take up is likely to be amongst the most technologically-literate.

Location-based services

It may seem rather sinister to get an SMS on your mobile phone saying 'Are you hungry?' as you approach the Big Bang Burger Bar. However, the manager of the Burger Bar will be pleased if it makes you decide to drop in.

At the moment location-based services only work when the user has already revealed their location, for instance by sending a postcode. However, the time will come, particularly with improvements in the networks, when providers and applications will be able to tell accurately where you are, beyond the cell you are in.

The US Federal Communications Commission (FCC) mandated in October 2001 that all cellular carriers must pinpoint calls to emergency services from mobile phones to within 100 metres for 67 per cent of calls and to within 300 metres for 95 per cent of calls (E911 or Enhanced 911). Unsurprisingly, compliance has not been universal in the US and is dependant on many things including new/upgraded handsets and the readiness of designated Public Safety Answering Points (PSAP).

E911 has prompted manufacturers to implement ways of complying with this mandate. The methods are typically based on triangulation between radio masts or use of the Global Positioning System (GPS).

Compliance with this mandate has resulted in forecasts that all networks will offer this accuracy in location-based services by 2006 and that businesses can expect corresponding growth in location-based services.

Subscribers to location-based services

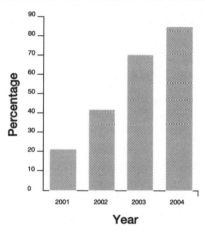

Of course, knowing where you are and offering you a burger is amusing, but the commercial opportunities offered by location-based services are more relevant.

Vehicle tracking
Many businesses need to manage their fleets of vehicles. Knowing where a car or truck actually is represents a big step forward in this. The 'tracking' device may be a mobile phone carried by the driver or a smart device held within the electronics of the vehicle itself.

Not only can businesses track where their vehicles are, but they can also determine how quickly they are travelling between locations. These will have benefits for businesses involved in:

- Vehicle hire
- Dispatch and delivery (see below)
- Taxiing

Shipment tracking
Shipping companies themselves are taking up M-Commerce. If you deal with shipments, this can help you to deliver your products to your customers. With this shipping companies can respond to your requests and to those of your customers for the location of shipments and their expected delivery time.

Directories
Location-based services are not just about being able to tell where employees are. A number of wireless directory services are now available, including the ability to tell users their nearest:

- Bank
- Fast-food outlet
- Hotel
- Golf course
- Car park
- Medical service (doctor, dentist, hospital or pharmacist)
- Underground station

Much of this is already available on the internet, where users need to say exactly where they are or plan to be. When a mobile device knows its location, provision of this information should be automatic, either as a response to a request or 'pushed' by businesses, like the Big Bang Burger Bar!

These services are of great interest to individuals, businesses and those who see individuals as potential consumers, should not only be aware that these services exist, but take steps to ensure that their business is included in the answer to a 'Where's my nearest . . .?' question.

Mapping, routing and traffic information
Mobile workers are, by definition, on the move. Facilities now exist to provide information on mobile phones or PDAs about journeys, either planned or while events are actually happening.

We all know how quickly traffic conditions can change. The most carefully constructed plan and route drawn from the internet can quickly be spoiled by an accident. The immediacy of mobile information will help to reduce dispatch or service delays, and indeed delays to planned meetings with customers.

Enthusiasm in the 1990s for mobile data services and WAP was short lived. Typically operators offered network services but left businesses to think how they were going to mobilise their applications. But a second generation is now coming on and they are ready to try again. They now work on GPRS and are more robust services.

Companies are now selling software that enables operators to offer businesses the ability to extend corporate applications to mobile devices by doing nothing more than dropping a three Megabyte file into their messaging or database server. Using this client installation, Java-based connectors call a user's MS-Exchange, Lotus Notes or database server and asks it to send specific information, such as e-mail, calendar or other database information. Typically operators will charge £10–25 a month.

Summary

Much of what M-Commerce offers business may be limited only by our imaginations. It is a world that is changing fast. We can today, however, look at what is presently available:

- Access to applications typically on our desks, e.g. agenda and address book
- Access to corporate systems including CRM
- Location-based services, such as:
  1 Vehicle tracking
  2 Shipment tracking
  3 Directories
  4 Mapping, routing and traffic information

A common characteristic of mobile applications is how they free workers from their normal workplace and so can deliver a faster and better service to clients.

Governments may not be commercial organisations in themselves but it is worth mentioning here the sort of initiatives they are doing or planning to do. It may inspire you for your own industry to see what sort of mechanisms are possible. The work is generally divided between enabling mobile citizens and enabling mobile employees. This is certainly equivalent to enabling mobile consumers and of course enabling mobile employees of businesses.

For citizens, typical government applications are access to government services such as:

- Traffic updates
- Tourist information
- Emergency alerts
- Mobile voting and referenda
- Public opinion and demographic surveys
- Tax and public services payments
- Emergency services reporting
- State lottery results
- News
- Weather forecasts
- Exam results
- Status checks on transactions

For their employees, there are initiatives to increase productivity such as access to the office and connectivity when on the move for things like intranet and e-mail, etc.

There are many specific applications to support mobile workers, such as:

- Law enforcement activities
- Fire and disaster emergency services
- Emergency and general medical services (e.g. blood donor contacts or test results in rural areas)
- Public-transport, especially real-time supports for drivers (e.g. changing traffic signals)
- Tax Collection

What M-Commerce means for individuals

A great deal has been said about the potential that M-Commerce offers business. Sadly, much of this remains to be realised and we must remain conscious that the introduction of WAP is hardly an encouraging forerunner.

Naturally, large amounts of the investment in M-Commerce to date has been made by those businesses which may clearly benefit, i.e. handset manufacturers and network operators, in other words, the enablers. Take up by the wider business community may seem slow but it is fast in the context of the speed with which the technology has developed.

It is worth considering the following memo from Western Union in the 1870s:

> *'This telephone has too many shortcomings to be seriously considered as a means of communication. The device is inherently of no value to us.'*

A substantial proportion of the adoption of mobile commerce by business has been in response to demand by individuals, either as employees or as consumers.

This consumer-led demand may be welcomed by marketing people. They are not faced with creating a new market for new products. Instead there is already an expectation amongst individuals that services exist, or will do in the future. It remains the responsibility of business to manage these expectations appropriately. It is certainly unusual for individuals to be more familiar with new technology than businesses are. We only have to look at the enthusiastic text messaging by teenagers on mobile phones to give us a clue about where the future lies.

NTT DoCoMo

We have so far refrained from mentioning proprietary products or services. However, iMODE, the service provided by the Japanese operator, NTT DoCoMo, has undoubtedly been a big success. We can learn much from it because it has paved the way for what may become available in the rest of the world.

iMODE is big in Japan because it is fast and cheap. It is accessed by mobile phones and users can:

- Send or receive e-mail
- Transfer funds between bank accounts, pay bills
- Execute stock trades
- Book plane tickets and make hotel reservations
- Find the nearest restaurant or hotel, with guides
- Get news on business, current affairs and sports events; weather forecasts
- Consult telephone directories
- Find city information
- Use dictionaries
- Access entertainment, e.g. games and horoscopes

iMODE is fast because it uses a W-CDMA (Wideband Code Division Multiple Access) network (similar to Europe's

GSM). It is relatively cheap because of NTT DoCoMo's charging structure, which consists of a monthly subscription fee and a charge for each packet of data sent or received. The average subscriber pays approximately $25 a month.

Above this, DoCoMo also charges approximately 9 per cent commission on transactions and enjoys an increased use of normal voice calls. We must not forget that the devices used to access iMODE are still phones and people will frequently call a business found in a directory.

A considerable percentage of M-Commerce applications development has been preoccupied with finding the 'killer application'. This application would be so important that individuals have to subscribe, in the same way that the Visicalc spreadsheet initially encouraged many people to buy PCs.

No clear 'killer application' has been identified yet, but it is understood that M-Commerce would die if network operators set their fees too high. NTT DoCoMo has avoided this.

The success of iMODE may have to do with the
characteristics of Japanese society, culture and geography but
we are already seeing the service being emulated, either in
full or in part, elsewhere in the world.

Users of M-Commerce

Unsurprisingly, it is the users of M-Commerce applications,
devices and networks, rather than specific businesses who
are the key to its success or failure. We can look at these users
from two points of view:

- Consumers
- Workers

Many users will be operating applications described
yesterday under 'business' because we are each of us
employees or consumers. As consumers, we take advantage
of the services and products that businesses offer. As
employees, we use mobile processes to further the
commercial interests of our businesses.

Consumers

These are the people who use services such as those offered
by iMODE. Consumers use M-Commerce to make their own
lives easier. It does not necessarily introduce new things that
we can do or change our lives. Nonetheless, it can change the
way we do things.

As we have seen before, M-Commerce offers a new channel of action. Now we can do things faster, more immediately and more easily.

There are features available to us on our mobile phones and PDAs that we can complete without mobile devices, such as:

- Buying things
- Finding things
- Banking (there is still no mobile device that will produce cash, although they can be used for non-cash purchases)
- Entertainment, e.g. games, music and gambling
- Communication, including SMS and chat services

However, it is attractive to use mobile devices because they make it easier.

As well as what the devices can do now (who would have guessed 50 years ago that our telephones could remind us of events?), convergence between them has meant that we have less to carry around with us. Having small and light devices is critical to the success of M-Commerce.

Perhaps it is useful to consider how M-Commerce changes life for the consumer.

Entertainment
We have never been short of sources of entertainment. Mobile devices may bring little in terms of new entertainment. We can still go to cinemas, concerts, sports events and casinos, read magazines, books and newspapers,

and play games. Possession of a mobile device is no substitute for the reality of these things and we should be wary of using them to lead vicarious lives.

Even the ability to play electronic games with another user is a new facility, previously unavailable to us except by using the internet, telephone land-lines or physical cables. Solo gaming, like Patience or Snake on mobile phones, is of course instantly available.

"ENJOYING THE ORIENT EXPRESS?"

The importance of M-Commerce is that it shows how these and other features can be used at very short notice and require little preparation. They can fill very brief periods when we might otherwise be unable to do things (e.g. when commuting).

This ability to fill 'niche' time playing games may seem trivial, but it indicates a key feature of the mobile world.

Time can now be seen as more valuable for us as individuals, than money. M-Commerce gives us the opportunity to use our time more effectively; time that could be wasted or unavailable.

M-Commerce is certainly not just for entertainment and the time can be used for all sorts of things, including making us more productive, by utilising diary management and composing or responding to e-mails. It is the sheer immediacy and availability of the mobile world that allows this.

Communication
It may be obvious that carrying a mobile phone helps us to communicate. The important difference to using land-lines is the 'anytime, anywhere' nature of M-Commerce.

As consumers, we expect to make more calls rather than receive them. Critically, we can call retailers and suppliers as and when we need to, rather than waiting to get home.

Furthermore, we can call sources of information we have found while on the move. Therefore, we can find out that a film is showing at our local cinema, and then call them to book tickets. This increased use of voice-calls is, after all, a valuable source of revenue to NTT DoCoMo with iMODE.

'Anytime, anywhere' is important when we need to talk to someone. Yet communication on the move is more than actually talking. It is more relevant to us as workers (e.g. e-mail), for entertainment (e.g. SMS, or as a medium for receiving information.

Information
PDAs, mobile phone features and the convergence of these devices mean that we can carry much of our personal

- Diary and calendar
- Address and telephone books
- Notepad
- To-do lists
- Calculator

information around with us, including the following:

Again, much of the value lies in being able to replace many of the things that we used to carry with a single device.

Obviously, having immediate access to our own personal information is useful, but the real commercial benefit to consumers lies in the external information that is now available.

The amount of actual information is perhaps limitless, or at least no less than what is available to us from home. The difference for a mobile consumer who wants information is the immediacy and the accessibility.

Vast amounts of the information to mobile services is

automated, such as train times, weather reports or directory services. Having that information to hand saves carrying around timetables, transistor radios (which can now be a part of mobile phones) and heavy copies of the Yellow Pages.

Those of us who have been frustrated when making voice-calls to organisations will welcome this degree of automation, purely because it can be quicker, more reliable and more efficient.

Equally, consumers can access personal information which they might not normally carry around, for instance bank balances or stocks and share prices.

A great deal of the information that consumers will look for on the move is, by virtue of its instant nature, location-based. As we saw yesterday, these services are clearly of benefit to the consumer or potential consumer.

The ability to determine a mobile user's location means that information sent or received can become much more focused. When someone asks for the nearest ATM (automated teller machine) they can receive exactly that, rather than a generalised non-specific list.

Business transactions
Buying goods is nothing new and there is little that can be done on a mobile device that cannot be done in person or from home.

M-Commerce, however, allows users to complete transactions while in transit and at their own convenience. Sadly, it is not yet possible to touch and feel potential purchases. As mobile devices develop you can view products, particularly when accessing the internet.

Transactions depend heavily upon the consumer already knowing the product they want. However, it remains possible to:

- Arrange a visit from a sales executive from the supplier organisation
- Place an order for delivery to any location

Workers

With the word 'workers' there is a distinction between individual consumers who use mobile devices or services for their own personal use, and individuals who use the devices in their work capacity, either as employees or employers.

Much of what was mentioned on Wednesday is relevant to this.

Most workers are distinguished from the consumer by their need to access corporate information internal to the company. Obviously, different types of worker will need different types of information, just as they do at work.

Mobile commerce means that these workers can be with their customers and still be able to access all the information that they need.

On a visit to a potential customer, a sales person can hear the customer's needs and problems and then immediately:

- Obtain relevant product or service information
- Obtain inventory and price lists
- Interrogate the customer's credit rating
- Place an order
- Execute the corresponding financial transaction
- Agree a delivery date, time and location

This may have taken a long time while each step in the process is executed, and similar steps could happen over telephone, e-mail and the internet. However, the mobility of individuals involved in the sales process means that they can actually talk with their customers face-to-face, and still have any necessary information to hand. The entire process is speeded up, allowing sales staff to see more people per day than they might do otherwise.

As with the internet, agreeing a sale is one of the first steps. It must be followed with a logistics exercise to physically deliver what has been bought. Here, staff involved in delivery (even if out-sourced to organisations such as FedEx, UPS or Consignia) can make productive use of M-Commerce tools.

Individuals in every industry may speak of how
M-Commerce can help their work. Logistics is no exception
and is an area where a great deal can be done, if only because
M-Commerce gives a greater knowledge of where people are.

Once a product is sold and delivered, it will need support
and maintenance and we can see how M-Commerce can help
visiting technical engineers. They can:

- Receive a call for support from a customer
- Identify the nearest available engineer
- Locate the customer and actually get there (which
  can be a major exercise in itself)
- Speak with the customer and identify the problem
- Review other similar problems there may have been
  with this product elsewhere
- Consult an expert and send pictures of the problem if
  necessary

In effect, the engineer can bring the entire experience and expertise of the company to the customer. All of this can be done in real time and there is no need for the worker to return to the company base.

Not only can workers access information about products they are selling or servicing, but all of them (including corporate executives) can access internal corporate systems while they are away from their base.

We looked at M-CRM systems yesterday. The information held there is of interest to all individual workers visiting customers. All corporate executives will also be interested in home base access while away from it. They can make valuable use of 'niche' time to access internal systems which are not necessarily directly related to individual customers. Equally, they can use mobile devices to check e-mail or relevant news in their wider industry.

Of course, all workers can now replicate many of the internal support systems and processes that they rely on when at their base. Applications such as diary management, contacts, e-mail, word processing and spreadsheets can be carried around by the individual worker.

Summary

M-Commerce offers little that is tangibly new to the individual worker or consumer. Importantly, though, it differs in its accessibility, relevance, convenience and speed. It recognises that workers and consumers are mobile, so need things in a different way. M-Commerce should be seen as a new channel for communication, not only enabling people to work both in the office and at home but *on the move*. It brings with it immediacy, both for the individual consumer who can find what they require more quickly and the business who can respond to consumer demands at greater speed.

The future – opportunities, risks and challenges

Throughout the week we have looked at the technology of M-Commerce, and how its applications can offer so much or can even transform the way we live our lives.

Today we will look at three of the most significant problems that are yet to be overcome satisfactorily:

- Privacy
- Security
- Funds transfer

Concerns about privacy and security are probably the most significant barriers to M-Commerce.

Privacy

We have examined location-based services earlier this week. Certainly, they are attractive to both businesses and to consumers. However, like many things, there is a flip side to the benefits; in this case a privacy cost.

We may want to find our nearest ATM or restaurant and will be pleased that our mobile phones know where we are in order to work this out. Equally, we may have misgivings about anyone being able to track our location, whereabouts and movements.

That 'anyone' could be Government organisations, law enforcement authorities or businesses. George Orwell's 1949 vision of the future, *Nineteen Eighty-Four*, becomes less whimsical in this context.

"IF YOU WANT A PICTURE OF THE FUTURE, IMAGINE A BOOT STAMPING ON A HUMAN FACE — FOREVER."

Source: George Orwell, Nineteen Eighty-Four *(1949), Pt 3, Ch 3.*

The vision for a world with M-Commerce is not so extreme. Nevertheless, a big difference between now and Nineteen Eighty-Four is that it is not just the state which could be watching, but also business.

Our society, and indeed commerce, often depends on trust. This leaves a burden of responsibility on businesses who want to operate in the M-Commerce area to build trust amongst consumers and employees.

It is perhaps inevitable that the introduction of new technology will inspire misgivings. Concern and anxiety about privacy may be exaggerated beyond what is actually possible and yet regardless of how logical or self-evident the reality may be, it is still up to businesses to inspire a feeling of trust amongst potentially affected users.

Generating and maintaining trust is an important issue in itself, and it is clear that consumers and employees have more faith in the integrity of larger and better established organisations and brands, however mistaken or deluded that trust may be.

Trust is not only affected by what individual businesses themselves could do with personal information, but also by the fear that they could sell it on. It is very valuable information meeting a wide interest. Commercial enterprises can use knowledge of their customers' whereabouts and movements, but there is equal, if not greater, concern about the ease with which personal data could fall into the hands of others.

The Wall Street Journal has reported that AT&T Wireless gets about 15,000 sub poenas annually for its records. Civil litigation is, of course, a potential concern to individuals but all businesses getting into M-Commerce should always be mindful of the legal risks they face.

There has always been a lag between the development of new technology and the implementation of a corresponding legal framework. It takes a long time to effect appropriate legislation. In the meantime, there is civil and criminal law in effect that can sue businesses, as the growing numbers of lawsuits in this area demonstrate.

The FCC's direction to network operators to be able to locate callers of emergency services will still be an attractive tool for law enforcement authorities (see Wednesday). Nevertheless, for them it may be a double-edged sword because criminals could use M-Commerce to construct false alibis just as easily as the police use it to determine whereabouts.

This typifies some of what civil liberty organisations fear. Although something is initially for worthwhile reasons, it does not take much for it to surreptitiously creep into a less noble domain.

Already we are seeing businesses hoarding personal information gathered from many sources, not necessarily for any particular reason, but 'just in case'.

However, it is not all gloom and doom – people can always switch off their mobile phones! It remains a tightrope that we have to walk; to balance our desire to use location-based services against our desire to preserve our privacy.

We do not even need to do anything as drastic as switching off our mobile phones, but can choose to opt out of location-based services. This may be a general exclusion, covering all applications and providers, or selective.

On the positive side, many people can achieve more privacy by choosing where they use a mobile phone which is impossible with a land-line at home or work.

Security

Apart from the individual's concerns about privacy, there are business concerns about security and this focuses on the money issue.

There is a danger with M-Commerce that not only can goods, money and services be criminally acquired, but businesses can lose whatever confidence they may have inspired in their customers. This in itself may be more damaging than losing 'money' and irrevocably more expensive to rebuild.

Many of the security concerns about E-Business (see Dave Howell, *E-Business in a week*, Hodder & Stoughton, 2002) still apply to M-Commerce, yet opinions are mixed over which represents the greater risk. In M-Commerce there is still a need to protect the security of any data while it is being transmitted from consumer to business, be it a bank or a retailer.

In the same way that individuals and businesses can go mobile, so can fraudsters. Security fears have already damaged online banks on the internet.

Confidence in M-Commerce could be hurt in the same way and by exactly the same mishaps. Therefore, an M-Commerce business could easily find itself tarnished by an E-Commerce fraud, even if unconnected with the mobile part of the business. Confidence is more about perception than pure logic, and the idea that sensitive data can literally be pulled out of the air can only arouse concern.

When looking at the security of M-Commerce businesses, we must look at each step on the value chain. The M-Commerce chain is no less complex than that in the terrestrial world and perhaps more so.

Most significant are the number of different organisations who may be involved in a transaction, for example:

- Consumer or worker
- Mobile network operator
- Land-line operator
- Your business

- Your suppliers
- Your bank
- The consumer's bank
- Payment operator and method

Each of these people or organisations will have their own security issues, but it is ultimately your business that will suffer from any security issues along the chain. All of the links between the various parties involved may be the weakest points in the M-Commerce chain, particularly when sensitive data is being passed. (These are points when information or instructions are passed from one individual or business to another, with all the potential for misunderstanding.)

We may soon be approaching overload in terms of the number of passwords we each have to remember and use in our lives. These and encryption or decryption are now used everywhere.

> **Secure server unknown**
>
> **Key exchange RSA, 1024-bit (high)**
>
> **Encryption: RCS, 40-bit key (low)**

Example of security message when connecting to WAP on a 2G mobile phone network

Apart from WAP, which has yet to truly catch on, SMS on GSM remains an important medium to transmit data. Although it is commonly used for sending personal messages, it is still used for financial and non-financial transactions. Its limit of 160 characters is a constraint, but this limit may be raised in the future.

Because GSM and SMS were evolutionary in their development and implementation, they do not necessarily have the security features required for financial transactions. GSM operators have sought to address this by using the Subscriber Information Module (SIM) card in mobile devices.

SIM cards hold a great deal of useful information about the subscriber, such as the mobile phone number, Personal Identification Number (PIN) and device security codes. Together with the allocation of private keys, all of this information can be used to answer a business' question of 'Who am I talking to?'

3G will meet many of the security shortcomings of 2G, but we should be mindful that it is constantly evolving technology and that change is in itself a security risk. IBM are getting involved in security on wireless technology and it is encouraging that they have recognised the area as a concern.

Biometrics
A key question for businesses in M-Commerce is 'Who am I talking to?'. This is no less so than when using the internet or even using voice over the telephone. Here the technology in mobile devices can serve to help.

Many of us now use a mobile phone or a PDA but it is very tempting not to use the password protection built into them, despite the amount of sensitive personal information they hold, including credit card numbers. Entering a password can be too much of a rigmarole to go through every time you use it. Over 250,000 mobile phones and PDAs are lost or stolen at airports every year and only 30 per cent of them ever make it back to their owners.

Biometrics is a way of using parts of the human body as a kind of permanent password. Technology is such that your mobile device will be able to identify you through differences in your retina, the shape of your hand, your fingerprints, your ears, your voice and a host of other physical characteristics.

Biometrics have an obvious place in banks, airports and high-security facilities but one of the most hotly pursued areas for biometrics is in hand-held devices and dozens of companies are working on this. They can make any device worthless to a would-be thief and almost eliminate fraudulent transactions. Voice and fingerprints are the two main technologies being considered by mobile manufacturers. Both of these are much more secure than PIN codes, which tend to be forgotten or written-down and then lost. They allow users to access banking information, voice mail, e-mail and other private records, without requiring them to enter a password.

When it comes to security, technology can help but it is not everything. Policies, procedures, education and training are also needed.

Funds transfer

Having addressed security concerns, we can move on to the 'How to get the goods paid for?' question.

Naturally, the ways of paying must reflect and be proportionate to the product or service that is being bought. We can already use credit and debit cards to buy things over the telephone or over the internet. There is no reason why this should not be the same when using M-Commerce, although it will still carry the same risks.

Payment for goods
Payment for goods can continue to be made by credit or debit card, providing it is acceptable to the retailer.

If the payment is for an acceptable amount, consumers can still pay per transaction. Re-keying account details while on the move is inherently risky, not to say inconvenient. It is much easier if a consumer already has an agreement with the supplier, including debit or credit card and delivery details, so that only a password or token need to be provided.

An important point here is that it is the supplying business, not the network operator, who can choose whether to accept or decline a transaction.

Subscription services

There are many services available to the mobile consumer, such as location-based services. These are services which can be best paid for with a monthly subscription to the provider of the information.

If there is no clear single provider, then the subscription cost can be raised by the network operator and recovered through the monthly bill.

Pre-pay mobile phones

There is clearly an issue in respect of pre-pay mobile phones; because there is no monthly bill involved, the network operator cannot levy an appropriate charge. Instead they can only deduct it from the value held on the SIM card of the phone itself.

Once people hold more money on their phones through buying more and larger pre-pay vouchers, it becomes apparent that network operators may be acting as banks themselves. Consequently they may find themselves constrained by the myriad of national and international banking regulations.

Quite apart from the administrative overheads involved in network operators providing this sort of payment service, the regulations themselves may deter them.

Micro-payments

In 1998, Nokia presented a service called Dial-a-Coke, where a consumer could dial a number on a vending machine and have the drink dispensed immediately.

This is an example of a mobile phone being used to make a small purchase and it is becoming increasingly widespread, particularly where mobile phone use is high, such as in Scandinavia.

The consumer enters into an agreement with the supplier agreeing either how the costs will be recovered, through the phone bill, a credit card or directly from a bank account. Like other payment methods, a prior agreement and password or token saves having to enter card details.

Micro-payments can be required frequently but only for small amounts and are of enormous benefit to the consumer

who no longer has to carry so much cash. There may still be an administrative overhead to suppliers, but they may also enjoy the benefit of not having to handle cash.

Payment for data services

Packet-switched networks have a big advantage in charging ability over their circuit-switched predecessors. With packet-switching we can measure the actual amount of data transmitted, rather than the time spent or flat rate. This allows more precise and granular charging in an environment where it can be difficult to predict, such as video and audio data.

Mobile wallet

Another payment solution is the mobile wallet, rather like the smart cards we already carry around such as those used at vending machines/canteens or even those smart cards which allow us access to restricted parts of the building. These need to be connected to or built into a mobile device and, once loaded, have stored value and can be used to pay for goods and services.

Public Policy Issues

We spoke on Wednesday about how governments can make use of M-Commerce facilities. But there are a wide variety of public policy issues that could affect the future of the industry and the commercial markets which operate within it. These issues include wireless safety, using mobile phones while driving, and the potential negative health aspects of cellphones. These concerns may have a vicarious negative impact on your business once it enters M-Commerce. So far, businesses have largely escaped government regulation in this area. However, Europe faces a higher degree of government regulation over the coming years led by the European Parliament. Much of this will mirror what has already been implemented or planned in the US.

There remains a lot of discussion about whether opt-in or opt-out policies should be put into place for location-based advertising. We can expect the FCC (Federal Communications Commission) in the US to set privacy rules for location-based services. In Europe most of the discussion has been centred on whether people should be able to opt-in or opt-out of participating. In November 2000 MEPs (Members of the European Parliament) voted to leave the decision to European member states and are also recommending that subscribers have the right to request that their names be removed from printed or electronic directories. The UK has backed the opt-out clause, which means users have to tick a box on a privacy statement to avoid being sent unsolicited messages. But Austria, Belgium, Denmark, Finland, Sweden and Germany are backing the opt-in, with users having to give their consent to receive unsolicited messages.

Driving and using your mobile phone

The use of mobile phones when you are driving has become a politically sensitive issue in the US and indeed elsewhere. New York State has already enacted a ban on the use of cellphones while driving and at least 39 other states are considering similar legislation.

So far the laws do not seem to have negatively affected the wireless industry. In fact, some service providers have actually supported the clause. It does provide manufacturers with the opportunity to sell hands-free devices.

In Europe, the Swedish authorities prefer a total ban on cellphone use by drivers of motorised vehicles. This does have a serious impact on mobile usage as US studies have suggested that up to 70–80 per cent of mobile calls originate from cars.

Although legislation on mobile usage while driving remains unclear in many countries, laws will start to be introduced that are likely to make it illegal.

Health related issues

To date there has been little conclusive evidence that mobile phone usage poses a health risk. Nonetheless, the subject is an emotive and politically charged one. Whilst it is not in dispute that mobile phones emit radio frequency radiation (RFR) from the antenna, the actual effect this has on the user is still uncertain.

One study funded by the UK government and undertaken by independent scientists warned people not to let children under the age of sixteen use mobile phones frequently because there are signs that the amount of radiation could have negative effects on developing brains. All these issues may remain unproved though civil court cases brought by members of the public are increasing. In 2002 US lawyers representing people who claimed mobile phones caused them brain tumours filed five cases asking for an excess of $9.75 billion in damages.

In August 2001, the European Commission set limits on the amount of electromagnetic radiation that can be emitted by mobile phones in Europe. Mobile devices currently marketed in Europe are well within these limits, but it will force companies to provide more information about their products. There is also public concern about the siting of radio masts. In response to this, in 2002, UK schools were given an effective veto on their siting.

As the number of mobile devices continues to grow, increasing legislation is inevitable. It may be that legislation reflects the maturity of the industry. Regulations may impede further growth for many vendors and for people attempting to use this technology for their commercial markets. The advancing technology – and 3G is expected to have many more base stations than GSM – must still be tempered by social responsibility, frequently through legislation.

Summary

There are many risks and issues presently visible around M-Commerce. Their number and importance is likely to grow as M-Commerce becomes more endemic and institutionalised.

These issues are not necessarily new to everyday business:

- Privacy
- Security
- Funds transfer

They must now be addressed from a new perspective because it is a whole new channel.

Exactly what needs to be done will depend on your own

industry. Much of it depends on businesses engendering and sustaining trust. Both consumers and employees must hold this trust.

The M-Commerce world

Trying to forecast the future in a mobile world involves a great deal of crystal-ball gazing. We are faced with a rapidly changing environment in terms of both the technology and what can be done with it.

This environment is certainly changing quickly in the M-Commerce area, but we should always be aware that other areas are also developing and the future is likely to see a great deal of convergence and overlap between areas, such as with the Global Positioning Satellite (GPS) service.

It may be convenient now to think of mobile phones and fixed-line communications as separate entities, but is it too far-fetched to think that all communication in the future will be carried through a chip embedded in our brains?

Perhaps the best we can do now is to look at what is available and treat promises and forecasts by suppliers and commentators with some degree of cynicism, particularly the further they look into the future.

Changes will not happen overnight, but businesses must still be prepared. Of course, changing our businesses for an uncertain future is hard. This may involve restricting ourselves to only tactical initiatives, rather than making large longer-term strategic investments. The inherent obsolescence risk in new technology may also steer us away from large investment.

What will the mobile community look like?
Because mobile devices can be used in so many different ways there will not be a single mobile community; there will be many. Individuals will use devices, such as mobile phones, to meet their own unique needs. Some people will use them to stay in touch with their children away at college, some want to have access and e-mail, some want to download streaming audio and video, some want to call home, some want to snap a little camera on to their phone, take a picture and send it to a friend.

Even before they start buying anything, there is as much variety in what people will want from their mobile phones as there are people. Naturally, this opens the doors to a correspondingly wide range of commercial opportunities.

It is a curious situation where people may be more comfortable sending e-mail or SMS via a mobile phone than walking up to people and actually having a conversation.

On one hand you can see this as a technology which enables people to exchange information on the move. On the other hand you can guess this is giving direct marketing companies the ability to spam you personally and wreak havoc on your personal life. These are the sort of perceptions that anybody entering M-Commerce should be aware of and sensitive to. The introduction of new technologies frequently introduces new paradoxes into our society. Better communications may come but ironically we may become more isolated, as we use the technology to moderate our surroundings more than ever before. We can choose to interact only with those elements that we really want to be in contact with.

What we know now

This week we have discussed changes that have happened or are imminent in:

- Networks
- Devices
- Applications for business
- Applications for consumers and individuals
- Risks and challenges

Access

At the 1995 Telecom Conference in Geneva, Larry Ellison (Chairman and CEO of Oracle) coined the phrase 'network computer' as a processor-only device with a browser but little in the way of applications or locally-held data. He renamed the network computer 'mobile phone' in 1999.

We can see how prevalent mobile phones have become, overtaking 'traditional' devices such as televisions and PCs.

The cost of mobile phones and their connection to a network has fallen sharply. They are now affordable to millions of people, which means millions of potential customers.

A new channel
The millions of people with mobile phones may not be new customers, but this is a new channel through which they can be contacted for sales, service, and marketing. Likewise, do not forget that many of the mobile phone users are not just consumers, but also employees who can use these devices to connect to their business applications and Personal Information Management systems.

This suggests that users today are less interested in using mobile devices to access the internet, than to access services that are local, about the local area or transactional. This further supports the view that the phrase 'Mobile Internet' is less applicable when it comes to M-Commerce; mobile, yes, but internet, no.

Anytime, anywhere
This catch-phrase for M-Commerce describes an important characteristic of the mobile world, but it is perhaps better reworded for businesses as 'Right thing, right time, right place in the right way'.

In M-Commerce, the time and place capability is largely made available to us by the technology; devices and networks. It remains up to businesses to determine how they will do the right thing in the right way.

M-Commerce can allow us to transform businesses. It can free us of many of the constraints under which we presently operate. It can certainly free us from our offices and workplaces and allow us to be closer to our customers. Equally, it can bring the customer closer to the business.

Standards

Much of the activity in M-Commerce has been in the absence of any dominant standard. Standards are now beginning to emerge, either through market pressure or as a result of agreement among regulators.

There are many active players in the market, insofar as it is possible to identify just one single market, and the network operators do not dominate the infrastructure.

The absence of standards may be a good thing in the short term; it will hopefully encourage the best practices to emerge and encourage competition and reduce prices.

In the longer term, however, there is a pressing need for global standards. Fragmentation of the market is not necessarily desirable and the 'anytime, anywhere' M-Commerce aspiration of a global market is already hampered by an inability of consumers to freely roam across the world.

What businesses should do now

It is of course difficult to determine a way forward in such a rapidly changing world. Much can be learned from current innovators, for instance NTT DoCoMo, and from having a basic understanding of the underlying technology and what it offers.

Businesses should not expect to suddenly be transformed by M-Commerce. However, M-Commerce may change features about how a business operates internally and the nature of its relationship with the external world, including customers, suppliers and partners.

Summary

The following are actions that businesses should take when they are considering M-Commerce:

- Examine their business processes, both internal and external, to identify where M-Commerce can relieve pressure points. Freeing staff should certainly reduce time spent and costs.
- Remove some existing processes or implement new ones in their place.

- Consider the issues including (but not confined to) security, privacy and payment mechanisms.
- Review internal IT systems that workers may wish to access while on the move. This should not be restricted to diaries and telephone lists, but should include other applications such as CRM and ERP.
- Assess how the workers will be affected as well as the consumers.
- Estimate the social impact of what they are doing. Mobile phones may be the most successful computer-based consumer product of the age. Yet very little is known about how mobile technology is changing the way people interact and cooperate with one another.

The main characteristics of M-Commerce will affect all of your thinking:

- It is always on
- It is everywhere
- It knows who and where consumers are
- It is easy for people to use
- It is secure

Exactly how it will affect you and what you have to do varies depending on the industry in which you operate. The following list shows in descending order the appropriateness of M-Commerce:

- Utilities
- Insurance
- Construction
- Oil and gas
- Pharmaceutical and healthcare
- Telecommunications
- Consumer products and retail
- Aviation
- Manufacturing
- Government
- Banking
- Entertainment and media
- Automotive
- Legal services

Exactly what you do will depend on the organisation you aspire to be and the benefits you are seeking, for example

- A mobile workforce, together with their visibility and management
- Reduce costs and time for services you provide
- Reduce travelling time
- Better control on assets and inventory
- Customer service and loyalty
- Improve efficiency and reduce costs

M-Commerce is no different from initiatives you may be taking already, but it is important that you are aware of this new channel for both your customers and your employees.

A fundamental aspect to bear in mind is your customers' perceptions of the mobile world. Customers now have greater than ever variety of choice of channel when it comes to buying goods and services: they may physically come to see you in person; speak to you on the telephone, visit your web-site, send you an e-mail or even write you a letter!

Participating as a customer in your business through mobile technology should, ideally, be the same as through any other channel. Customers will have become familiar with how to do things. They will expect to see this preserved no matter what channel they are using. Making things faster and more convenient is certainly welcome. But if you change any of the fundamental steps in the process you must expect resistance unless you accompany it with education and demonstrate the incentives.

If you have migrated your business to the internet and E-Commerce that does not mean that migration to M-Commerce will be simple. Your businesses may have learned much from having done this, but it may even have a negative risk when introducing M-Commerce. Your staff and your customers may have an expectation that E- and M- are the same.

The disappointing introduction of the first version of WAP has done little to encourage acceptance of M-. It only serves to reinforce a view that we are now looking at the mobile internet and that it is not much good!

Careful, customer-focused introduction of new initiatives is necessary for their acceptance and we must be wary of the situation being thought of as 'technology in search of a problem' rather than the other way round.